Astronomy: Space Systems

Reader

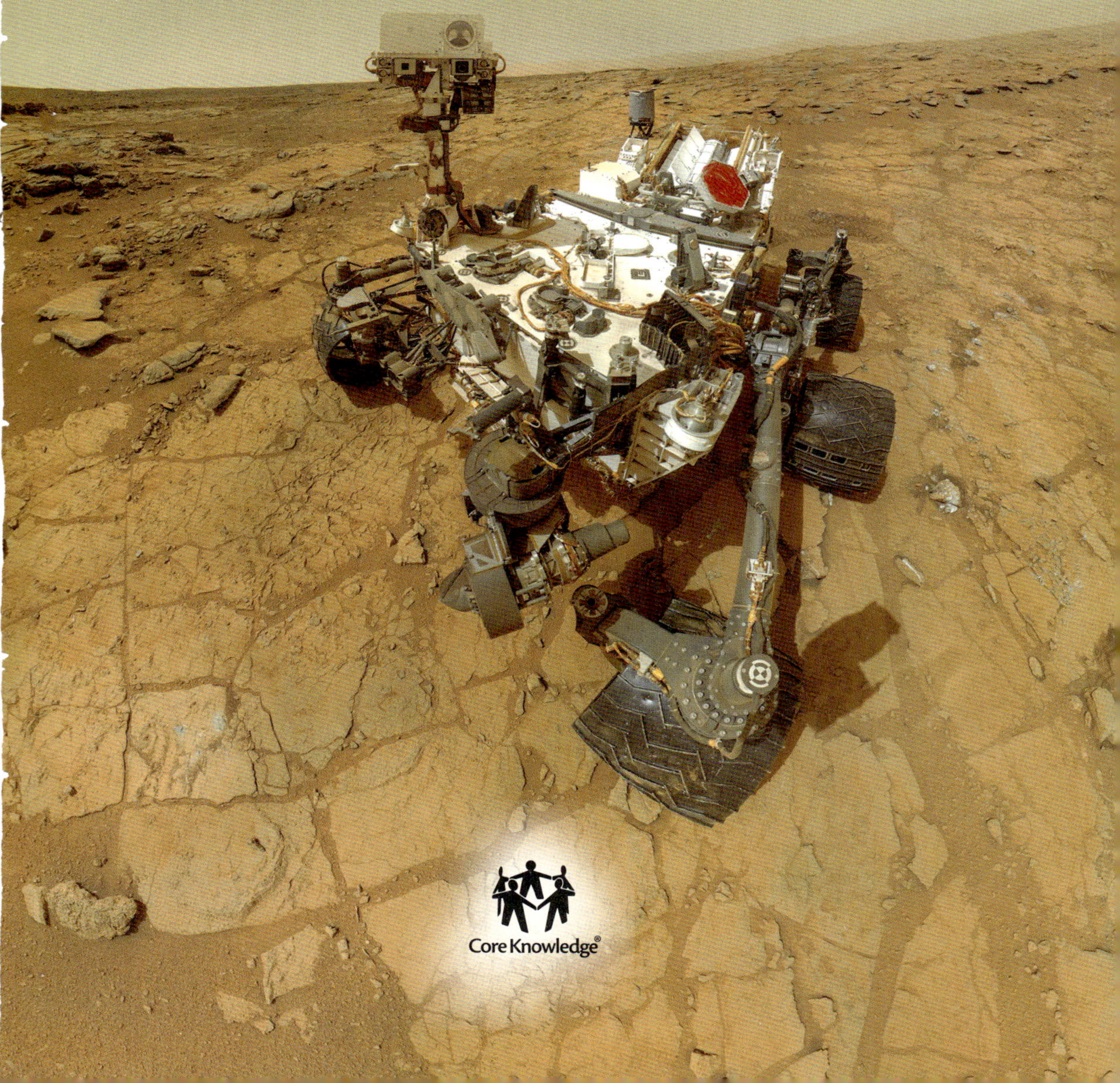

Core Knowledge®

Copyright © 2019 Core Knowledge Foundation
www.coreknowledge.org

All Rights Reserved.

Core Knowledge®, Core Knowledge Curriculum Series™,
Core Knowledge Science™, and CKSci™ are trademarks
of the Core Knowledge Foundation.

Trademarks and trade names are shown in this book
strictly for illustrative and educational purposes and are
the property of their respective owners. References herein
should not be regarded as affecting the validity of said
trademarks and trade names.

Printed in Canada

ISBN: 978-1-68380-554-0

Astronomy: Space Systems

Table of Contents

Chapter 1	**The Solar System**	1
Chapter 2	**The Vastness of Space**	7
Chapter 3	**Earth's Movement**	15
Chapter 4	**The Moon from Earth**	19
Chapter 5	**Brightness of Stars**	23
Chapter 6	**Constellations**	27
Chapter 7	**Gravity**	31
Chapter 8	**Women and Studying Space**	37
Glossary		43

The Solar System

Chapter 1

When you notice the sun, you're observing one part of our **solar system**. Earth is a part of this system, as are other planets and other objects that move around our sun.

A solar system includes at least one star and the objects that travel around it. Some of these objects are planets, large spheres that move around the star. Other objects in a solar system include moons, asteroids, comets, and particles of rock and dust. The path an object follows around another object in space is called its **orbit**. Objects are held in orbit by the force of gravity.

Our solar system has one star—the sun—and eight planets. Other solar systems have more than one star at their center.

Big Question
What is the solar system?

Vocabulary

solar system, n. a system of objects in space that includes at least one star, planets, their moons, asteroids, comets, and other space debris

orbit, n. the oval-shaped path an object follows as it revolves around another object in space (**v.** to revolve around another object)

Artists illustrate the solar system to show its main objects and their orbits. But they can't all actually be photographed together. They are millions of miles apart.

Our Solar System Contains Some Rocky Planets

When our solar system formed over four billion years ago, only rocky materials could survive the heat near the sun. Eventually, these rocky materials formed planets. The four closest planets to the sun are made up mainly of rock and metal. Each has a solid surface and is known as a terrestrial planet.

Mercury: Mercury is the smallest of the terrestrial planets and the one that is closest to the sun. How close? It is approximately 35,600,000 miles away! Mercury is only slightly larger than Earth's moon, and it orbits the sun once every eighty-eight days. Its surface is very hot.

An instrument aboard NASA's *Messenger* spacecraft made this image of Mercury's surface.

Venus: Venus may not be the closest planet to the sun, but it is the hottest! That is because Venus has a very dense atmosphere. Sunlight passes through Venus's atmosphere and is reflected off the surface. But then it is reflected back to the surface by the thick atmosphere. Venus is almost 67 million miles from the sun. It is about the same size as Earth and has many of the same features, including mountains and volcanoes. At one time, Venus may have even had a shallow ocean.

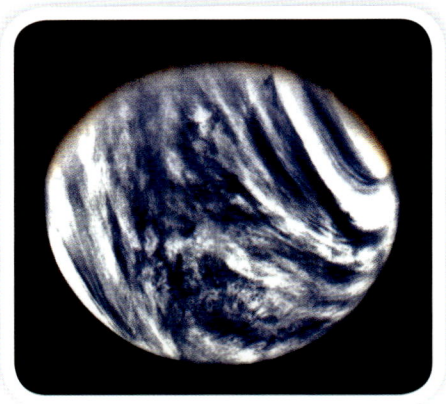

This photo from the *Mariner 10* spacecraft shows clouds swirling above Venus's surface.

Earth: Earth, the third planet from the sun, is the only planet known to support life. Earth's temperature, the amount of water on and below its surface, and the mix of gases in its atmosphere all make life possible. Earth also has its own moon. In fact, it is the only planet with just one moon. Mercury and Venus do not have moons, and the remaining five planets have two or more.

What features of Earth do you notice when you look at this image taken from space?

Mars: Mars, the fourth planet from the sun, is about half the size of Earth. It has a thin atmosphere and is very cold, with an average temperature of −81°F. The surface of Mars is dotted with mountains, craters, canyons, and volcanoes. There is no evidence, now, that life exists on Mars. But evidence of water on Mars has led scientists to consider that it may have supported life at one time.

Mars is sometimes called the "red planet" because large amounts of iron in the soil give it a reddish color.

Our Solar System Contains Some Gaseous Planets

The other four planets in our solar system are classified as either ice giants or gas giants. These very large planets do not have a definite surface. Instead of a rocky ground, they are mostly made up of gases, and only their small cores at the very center are solid.

Jupiter: Jupiter is the largest planet in our solar system. Jupiter also has seventy-nine different moons that we know of! Most people identify Jupiter by its colorful surface and the presence of its Great Red Spot. This spot is actually a huge storm that has been occurring for hundreds of years. The gases in Jupiter's atmosphere are mostly hydrogen and helium, the same elements that make up the sun.

Jupiter is eleven times the width of Earth.

Saturn: Saturn is nearly a billion miles from the sun. Like Jupiter, it is a gas giant that is made up mainly of hydrogen and helium. Saturn is most famous for its rings. Saturn has over fifty moons, and some of them may have the ability to support life.

Saturn's rings are made up of chunks of rock and ice.

Uranus: Uranus is one of the two ice giants. It is made up mostly of water, methane gas, and other materials. Methane is what gives Uranus its blue-green color. Uranus is a very cold and windy planet. Wind speeds can reach up to 560 miles per hour! Like Saturn, Uranus has rings. One characteristic that makes Uranus unique is that it tilts almost completely on its side as it orbits the sun.

Uranus as seen from *Voyager 2*. What gives Uranus its blue-green color?

Neptune: The other ice giant is Neptune, which is also the farthest planet from the sun. Neptune is very dark, windy, and cold. Its winds can reach up to 1,200 miles per hour. Like Uranus and the gas giants, Neptune does not have a solid surface. Its structure is similar to Uranus, with a swirling mixture of water, methane gas, and other materials. Some scientists think there may be a hot ocean deep under Neptune's clouds of gases.

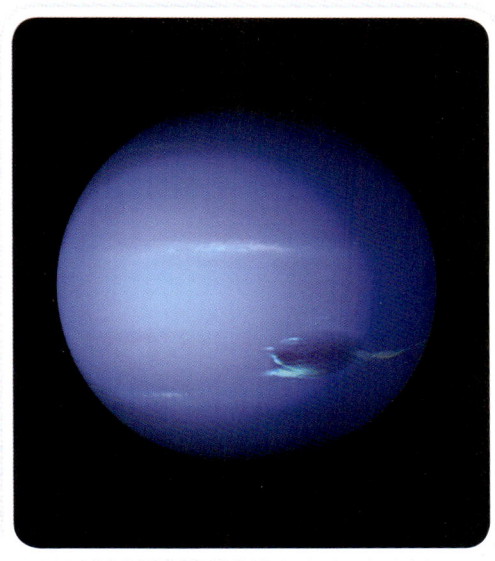

Why are there no rocky features to be seen on Neptune's surface?

Our Solar System Contains Other Space Objects

The sun and planets and their moons are major parts of the solar system. But the solar system contains other objects, too. Some of these objects are dwarf planets. A dwarf planet has some, but not all, of the characteristics of regular planets. Pluto is one example of a dwarf planet. Scientists have identified and named five dwarf planets. They are farther from the sun than Neptune.

Asteroids, meteors, and comets are also part of our solar system, all orbiting the sun and interacting with the planets. These objects range in size from small particles to large bodies that are hundreds of miles across. Comets contain ice and dust. When they are close to the sun, they form a "tail" of gas and other material.

A NASA spacecraft called *New Horizons* captured this photo of Pluto's icy mountains and frozen plains.

The Vastness of Space

Chapter 2

Every star may have its own solar system. Our solar system plus millions of others form a much larger system called a **galaxy**. Galaxies are collections of dust, gas, stars, and their orbiting objects. One galaxy may have hundreds of billions of stars. And there may be as many as 100 billion galaxies in all of space.

How do we know about objects in space? Scientists and engineers have developed telescopes and spacecrafts that explore deep space. For example, in 1977, NASA launched the *Voyager 1* and *Voyager 2* spacecrafts to travel past the edge of our solar system and collect data along the journey. Together, they continue to explore beyond the edge of our solar system, sending information back to Earth.

Big Question
What is the universe, and what are galaxies?

Vocabulary
galaxy, n. a collection of stars and their solar systems, dust, and gas

The *Voyager* probes took pictures of planets as they passed them and sent the photos back to Earth via radio signals.

7

Our Solar System Is Part of the Milky Way Galaxy

Our solar system is part of a galaxy known as the Milky Way. The Milky Way is a large, spiral-shaped galaxy that likely contains billions of other solar systems. Our solar system is located on one of the Milky Way's outer arms. All of the stars that you can see in the night sky are part of the Milky Way. Just as objects orbit the sun, our solar system revolves around the center of the Milky Way.

No person or spacecraft has ever traveled outside of the Milky Way to look back and take a picture of it. Scientists have pieced together clues from telescope images and other data to determine the Milky Way's shape and what it is made of.

Think about why the Milky Way galaxy might be named the way it is.

Galaxies Are Part of the Universe

Objects in space move and interact as systems. Scientists study systems and their parts to better understand and explain what they are made of and how they work. The sun, Earth, and other planets are part of a solar system. Our solar system is part of the Milky Way galaxy. The Milky Way galaxy is one galaxy in a very vast **universe**. The universe contains everything in space, including all forms of matter and energy.

Vocabulary

universe, n. all of the existing matter and energy in space

Over the years, scientists have collected data about the universe. They estimate that it is approximately 13.77 billion years old. It may have begun as a very small ball of hot, dense matter that suddenly exploded. It continues to spread out. How do scientists know this? By observing that galaxies are getting farther apart.

This image from the Hubble Space Telescope captured approximately fifteen thousand of the galaxies in the universe.

People Use Technology to Explore Space

Scientists have been using technology to observe, measure, and explore space for many years now. Over time, people have developed many tools for space exploration, including rovers, probes, satellites, and telescopes. Humans control these instruments from Earth.

Rovers: A rover is a probe that can travel after it has landed. It's a robotic vehicle remotely controlled to study the surface of a planet or moon. Rovers gather samples and collect data. They send information back to scientists on Earth. A rover named *Curiosity* landed on Mars in 2012. One of its purposes is to help scientists determine whether Mars might host tiny life-forms called microbes. *Curiosity* analyzes soil and rock samples. Two earlier rovers, *Spirit* and *Opportunity*, landed on Mars in 2004. They were intended to operate for about three months. But *Spirit* continued to collect data for ten years, and *Opportunity* operated for fourteen years. They found evidence that Mars used to have a watery surface and warmer climate.

In addition to transmitting photos back to Earth, rovers have instruments that determine the chemical makeup of material samples.

Probes: A probe is a kind of spacecraft that is launched from Earth on a rocket and sends information back to Earth via radio signals. Some probes land on other planets or moons and send back data from their landing site. Other probes collect data about the objects they are studying as they fly by without landing. The *Voyager* probes have traveled, and are still traveling, farther than any other spacecraft.

Satellites: A satellite is either a natural or human-made object that orbits another object. Human-made satellites are launched from Earth on rockets and guided into orbit around the objects they are sent to study. Satellites gather data, including photos, and transmit them back to Earth. Satellites take photos of planets, asteroids, and other objects of the universe. They help scientists study parts of the universe that are difficult to explore.

Telescopes: Telescopes are instruments that show faraway objects in greater detail. Telescopes may consist of lenses and mirrors that focus light to magnify images. Or they may collect radio waves from distant objects and use them to construct pictures of the objects. The Hubble Space Telescope was launched into space in 1990. Hubble collects light. It has revealed the formation of stars and images of distant galaxies.

The Hubble Space Telescope took this image of clouds of dust and gas. Scientists call this the Pillars of Creation. They were released during the formation of new stars inside the Eagle Nebula.

Humans Explore Space

John Glenn climbs into the *Friendship 7* spacecraft.

Probes, rovers, satellites, and telescopes are spacecrafts that travel without humans on board. But there have also been spacecraft that have taken human crews into space. Women and men from many different nations have orbited Earth, and some even visited the moon.

On May 5, 1961, Alan Shepard became the first American to travel to space. He flew aboard the *Freedom 7* spacecraft, and the flight lasted about fifteen minutes. This mission gave scientists valuable information about how the human body behaves during space travel. It also helped them identify and fix problems with space flight equipment and procedures.

Just a few months later, John Glenn became the first American to completely orbit Earth. He circled the Earth three times in just under five hours inside an aircraft known as *Friendship 7*.

Words to Know

A *crew* is the person or group of people who carry out duties aboard a vessel.

Uncrewed spacecrafts do not carry people.

Crewed missions have people onboard.

Humans first landed on the moon on July 20, 1969, during the Apollo 11 mission. American Astronauts Edwin "Buzz" Aldrin and Neil Armstrong landed their lunar module, the *Eagle*, on the moon's surface in an area known as the Sea of Tranquility. They spent two hours on the moon. During this time, they collected rock and soil samples. They also conducted experiments to learn more about the moon's environment.

Space Station: Did you know that there is a place in space where humans can live for days, weeks, or even months? The International Space Station (ISS) is a moving laboratory in space. It orbits Earth once every ninety minutes. Scientists from around the world arrive at the ISS via spacecraft. Then they live there while they conduct experiments in physical science, Earth science, and biology. If you go outside on a clear night, you may be able to see the ISS as it travels across the sky!

Astronauts live and work inside the tube-shaped modules of the station for months at a time. The section that the arrow points to in this picture is about the size of a school bus.

Humans May Travel to Mars and Beyond

So far, humans have not traveled any farther from Earth than the moon. But scientists are working to develop ways to explore deeper parts of space, with the goal of someday landing on Mars. One of the most important developments is the design of new vehicles, such as NASA's *Orion* spacecraft. *Orion*'s first mission will be uncrewed, and it will travel thousands of miles past the moon. This is farther than any spacecraft built for human space exploration has ever gone. Eventually, scientists are hoping that this spacecraft will take humans safely to Mars.

Scientists also have plans to use another rover to explore a region of Mars that they think may have been able to support life at one time. The rover will drill to collect rock and soil samples deep below the planet's surface. Then it will store these samples for scientists to study at a later time. This rover will also help gather information that scientists can use when planning the first human space flight to the red planet.

The *Mars 2020* rover is designed to collect samples. A future mission might return those samples to Earth.

Earth's Movement

Chapter 3

Focus on where you are right now as you read this. Are you moving? Yes, you are, even if you are holding very still. Planet Earth is moving, so you are moving with it. Because of your tiny size compared to Earth's size, you cannot feel Earth's movement through space. But relative to the sun and other objects in space, Earth is in constant motion. Although you cannot feel Earth's motion, you can observe clues about it by looking to the sky.

Big Question

How does Earth move in space?

Think about riding on a park merry-go-round. On one side of the merry-go-round are trees and a bench. You see them speed past as you spin. Around you go, and in the other direction you see a swing set and climbing bars pass by. But are the trees and swing set really passing by your field of vision? No, it is you and the merry-go-round that are in motion. This is what happens when the sun and stars seem to move across the sky in a day or a night. Earth is spinning, so you move beneath the section of the sky where those objects are visible.

The merry-go-round rotates around a center point. What the rider can see changes as she rotates with the merry-go-round.

15

Earth Rotates on Its Axis

A merry-go-round is a flat shape that spins. Earth is a sphere, the shape of a ball. Earth rotates around its **axis**. An axis is an imaginary line through the center of an object. The axis is a fixed point of reference.

> **Vocabulary**
>
> **axis, n.** an imaginary line through the center of an object that is a fixed point of reference

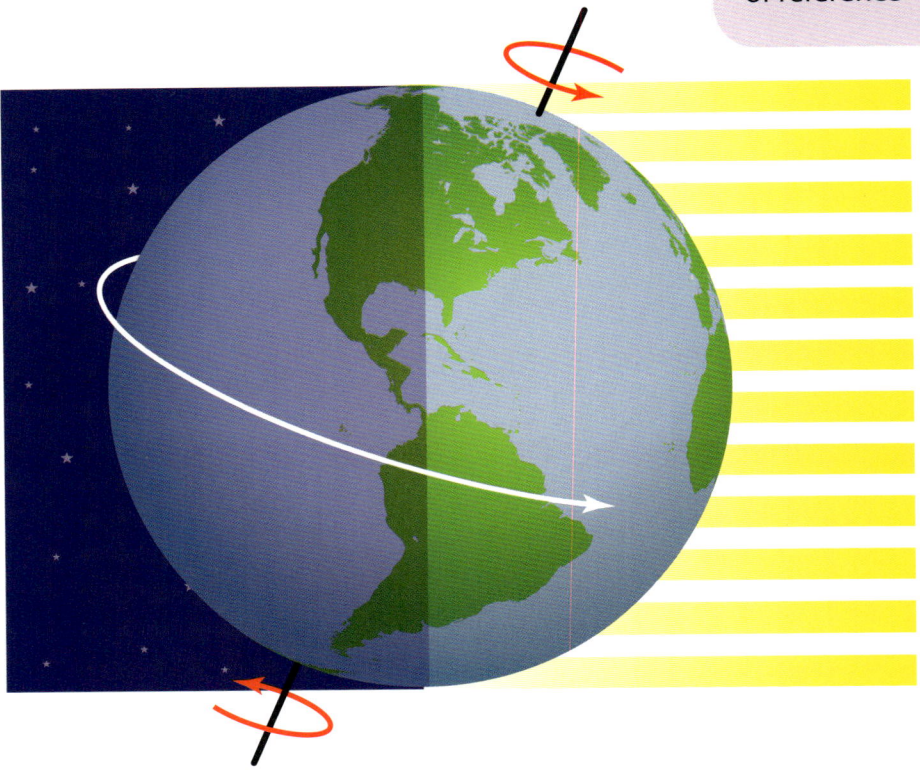

What you see in the sky depends on where you are on Earth's spherical surface. Two people on opposite sides of Earth's surface see different things when they look to the sky at the same time. One sees night, and one sees day. And what you see in the sky changes. It depends on which direction your location on Earth is facing at any given time. Earth makes one full rotation around its axis every twenty-four hours, one day.

16

Throughout the course of a day, the sun appears to move across the sky. When the sun is not visible at night, patterns of stars become visible. They, too, appear to move across the sky. But remember that this is apparent movement. Earth's motion changes where these objects appear in the sky.

If you observe the apparent movement of the sun and the stars, day after day and night after night, a very clear pattern emerges. They move across the sky from east to west. Earth spins from west to east. As you, at a point on Earth's surface, spin past objects in the sky, they appear to you to be moving in the direction opposite of what you are actually moving.

Shadows change throughout the day as the position of the sun changes. This is evidence that there is movement in the system that includes Earth and the sun.

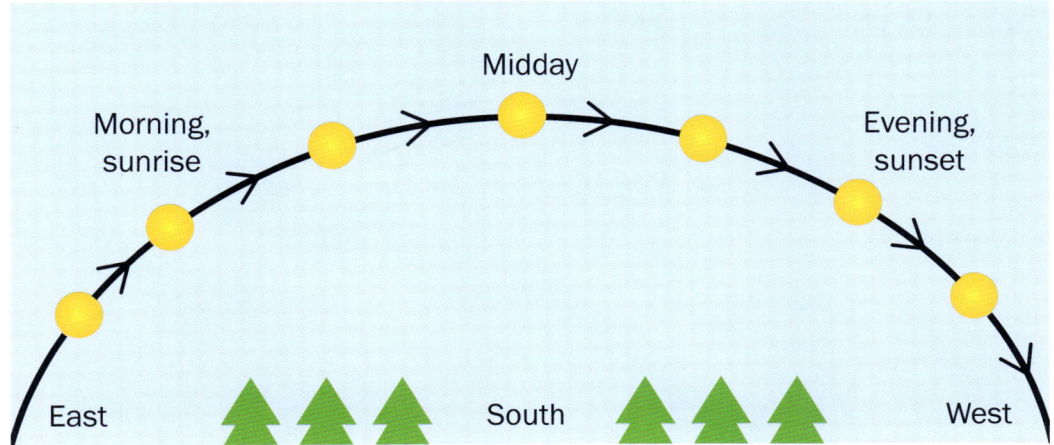

The sun's path across the sky is always from east to west. But the shape of its arc depends on the season and location. From most of the United States, in the Northern Hemisphere, the sun's path across the sky appears generally like this.

17

Earth Revolves Around the Sun

In a single day or night, you can see evidence of Earth's rotation. This is evidence that Earth spins on its axis. But you can also see evidence that Earth is moving in an orbit around the sun. As Earth moves in its orbit, we can see different constellations.

Constellations appear to move across the sky every night. This is evidence that Earth spins around its axis. In addition, visible constellations change with the seasons. This is evidence that Earth also orbits the sun.

In Earth's orbit around the sun, the nighttime side faces different regions of space, with different constellations, during different seasons of the year. Stars that are visible during summer nights are on the daytime side of Earth during the winter, so they are not visible then.

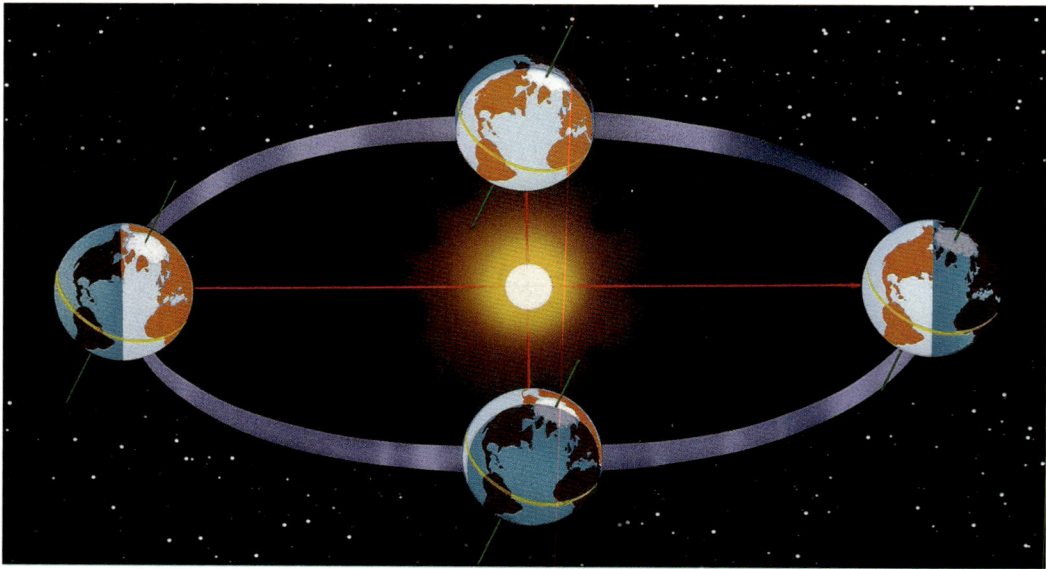

During different seasons, Earth's nighttime side faces different regions of outer space. So, in different seasons, we see different sets of stars in the night sky.

The Moon from Earth

Chapter 4

The sun, Earth, and the moon are parts of a system. The three objects interact in space, pulling on each other with the force of gravity. Like Earth, the moon is a sphere that rotates on an axis. The moon orbits Earth as Earth orbits the sun. Its revolution around Earth takes about twenty-nine days.

Big Question

What are moon phases and eclipses?

When we see the moon cross the sky from east to west in a single night, the apparent movement is a result of Earth's rotation. But other changes in the moon's appearance are also visible. On some nights the moon appears as a complete bright circle. On others it appears as a narrow, curved sliver. These differences result from the relative positions of the sun, the moon, and Earth.

Neither Earth nor the moon gives off its own light. These bodies are only visible because their surfaces reflect, or bounce back, light from the sun. Like Earth, one half of the moon's spherical shape is always bathed in sunlight. However, because of the changing positions of the moon and Earth, we can't always see the portion of the moon that is fully lit. This causes the moon to appear different to us at different times during its orbit.

On one day during the moon's orbit, the side of the moon facing Earth is fully lit by the sun. It appears as a complete circle and can be seen both at night and during daylight hours.

The Phases of the Moon

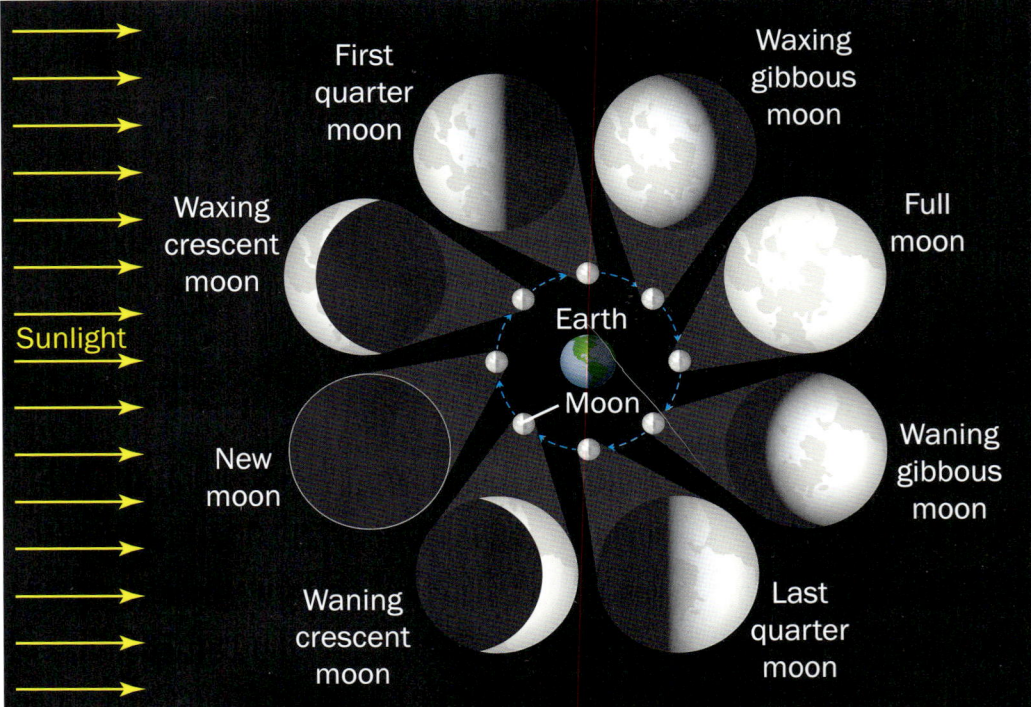

As the moon orbits Earth and Earth orbits the sun, their positions in relation to each other cause the moon's shape to seem to change in the night sky. These changes are called **moon phases**.

One half of the moon is always lit by the sun. But not all the sunlight that hits that half is reflected to Earth. A full moon occurs when the whole side of the moon facing Earth is sunlit. A new moon phase occurs when little or none of the sunlight that hits the moon is reflected down to Earth. All the phases of the moon occur over and over in a predictable pattern, repeating with each orbit.

Vocabulary

moon phase, n. a stage in the repeating, predictable pattern of change in the moon's appearance from Earth

Earth Can Block Sunlight from Reaching the Moon

Occasionally the appearance of the moon changes for a short time for reasons other than the current moon phase. Earth and the moon are both in motion around the sun. Sometimes the three objects line up, with Earth directly between the sun and the moon. When this happens, the moon passes through Earth's shadow. This event is called a **lunar eclipse**.

During a total lunar eclipse, Earth's shadow blocks all the light reaching the moon. When viewed from Earth, the moon appears to be a deep orange color as Earth's shadow passes over it. The color is a result of Earth's atmosphere filtering the indirect sunlight that passes through it to reach the moon.

Vocabulary

lunar eclipse, n. an event during which the moon passes directly behind Earth and into its shadow

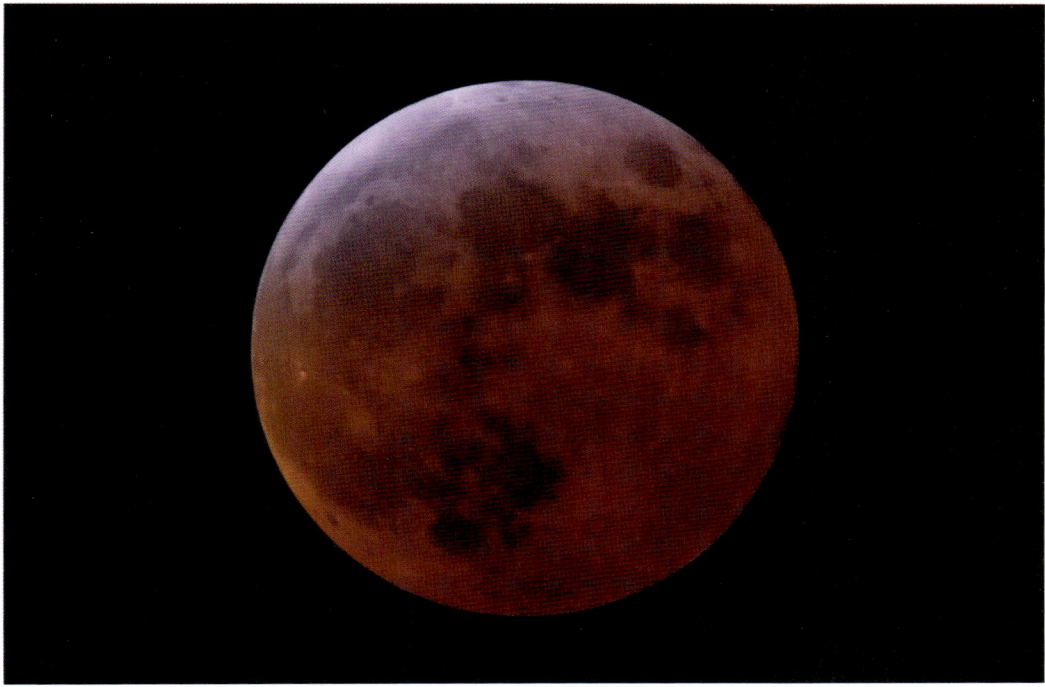

Total lunar eclipses occur when the moon is full.

The Moon Can Block Sunlight from Reaching Earth

The sun, the moon, and Earth can also line up with the moon directly between the sun and Earth. In this alignment, the moon blocks sunlight and casts a shadow on Earth's surface. This event is called a **solar eclipse**.

> **Vocabulary**
>
> **solar eclipse, n.** an event during which the moon's shadow blocks all or some of the sun's light from reaching Earth

The perfect alignment during which the moon is directly between the sun and a place on Earth's surface only lasts a few minutes in any given place. During that period of total solar eclipse, part of the Earth is eerily dark, and a halo of sunlight around the sun called a corona is visible.

The sun's corona should only be viewed through a special filter during an eclipse.

Satellites in orbit can capture photos of the moon's shadow passing across Earth's surface during a total solar eclipse.

Brightness of Stars

Chapter 5

When you think about stars, what comes to mind? You may think of the tiny points of light that you see in the night sky. Have you ever thought about which star is closest to Earth? You might also think of the sun, even though it looks bigger and brighter than other stars. Although the stars in the night sky seem very different from the sun, they are all stars, and they all have certain characteristics in common.

Big Question

Why do some stars appear brighter than others?

Stars are space objects that give off their own heat and light.

But why do some stars seem brighter than others? Is it because some are closer to us than others? Or do some actually give off more light than others?

Vocabulary

star, n. a space object that gives off its own heat and light

The sun only seems so much brighter than other stars because it is so much closer to Earth.

23

Brightness Changes During a Star's Life Cycle

One reason that stars are of different brightness is that stars change over time. Stars differ during these changes by age, size, and temperature. They all begin as clouds of gas and dust, but they change over the course of billions of years. Stars follow predictable patterns of change as their matter and energy proceed through the reactions that make them glow. Eventually, all stars "die."

Stars progress through called a life cycle, and their brightness changes, too.

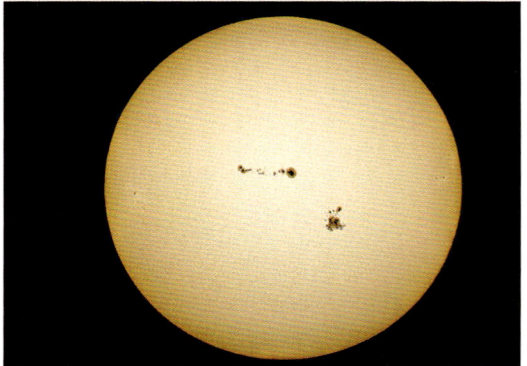

The sun is classified as a yellow dwarf. It is a medium-sized star. Yellow dwarf stars last approximately ten billion years.

Most stars in the Milky Way are classified as red dwarfs. They are smaller and cooler than yellow dwarfs. Many are so dim they can't be spotted without powerful telescopes.

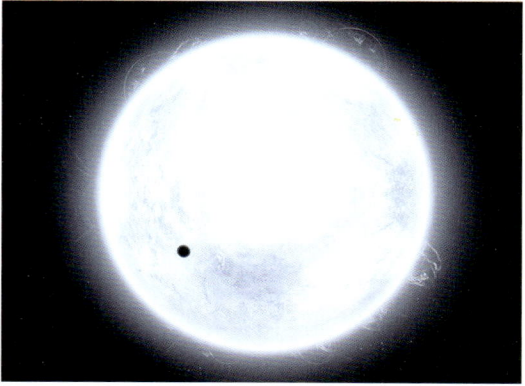

The star named Rigel is a blue giant, twenty-five times larger than the sun. Blue giants use up their energy relatively quickly and become cooler red giants.

Brightness Is Both Apparent and Absolute

Imagine that a ship is approaching a lighthouse at night. From far away, the light appears small and dim to the ship's crew. However, as the ship gets closer to the lighthouse, the light apparently seems bigger and brighter. The light's brightness has not changed.

A star's **apparent brightness** is how bright the star looks when viewed from Earth.

But stars of different ages, sizes, and temperatures actually give off different amounts of light. Some are truly brighter than others. Scientists have established a measurement called **absolute brightness** that identifies how bright stars are when viewed from the same distance. The standard distance used for this comparison of stars is 32.6 **light-years**. A light-year is the distance that light travels in one Earth year. Light travels very fast. One light-year equals approximately six trillion miles!

> **Vocabulary**
>
> **apparent brightness, n.** how bright a star appears compared with other stars when all are viewed from Earth
>
> **absolute brightness, n.** how bright a star is from a standard distance of 32.6 light-years
>
> **light-year, n.** the distance that light travels in one Earth year

A star with a high degree of absolute brightness can appear dimmer than a star with lower absolute brightness if the brighter star is much farther away. Many of the tiniest points of light in this telescope photo have greater absolute brightness than the stars that look brighter only because they are closer to Earth.

What are the brightest and closest stars to Earth? The sun, of course, is the closest. It is about 93 million miles from Earth, or 8.3 light-minutes. The sun's apparent brightness is so great that no other stars are visible from Earth's surface facing the direction of the sun.

Proxima Centauri is the second closest star to Earth. It is 4.2 light-years away. If you look at Proxima Centauri through a telescope, the light you see was given off by the star over four years ago!

The light that Proxima Centauri is producing right this moment will not be visible from Earth for over four years from now.

The star that appears brightest to the unaided eye in the Northern Hemisphere is called Sirius, or the Dog Star. Scientists know Sirius is an extremely large, bright star. It appears brighter than Proxima Centauri even though Sirius is almost nine light-years away. What kinds of things were you probably doing when the light now visible from Sirius began traveling toward Earth?

The brightest star in this night sky photo is Sirius. Sirius has an absolute brightness greater than that of the sun. Its apparent brightness is much less than that of the sun.

Constellations

Chapter 6

On any given night, thousands of stars are visible in the night sky. Some appear to be slightly bigger, smaller, brighter, or dimmer than others.

Big Question
What are constellations?

People from cultures all over the world have gazed at the night sky for centuries. People have imagined different patterns stars seem to make, like connect-the-dot pictures. Names and stories about stars and patterns of stars have helped people differentiate and remember stars. For example, the ancient Greeks identified many star patterns and named them after Greek gods, goddesses, and mythical creatures.

Vocabulary
constellation, n. a group of stars that forms a recognizable pattern in the sky

A **constellation** is a group of stars that forms a recognizable pattern in the sky. Constellations are things of the human imagination. But astronomers often use them as a way of communicating where a star is found in the sky. A scientist may say, "The bright star Sirius is found in the constellation Canis Major."

Imagine Canis Major as a connect-the-dots picture of a dog. Sirius is the star at its collar. The star at its nose is called Adhara. The star at its tail is called Aludra.

Different Constellations Are Visible at Different Times and Places

Because Earth orbits the sun, we see different parts of the night sky at different times of year. Also, your location on Earth and the time determine what region of space you are facing.

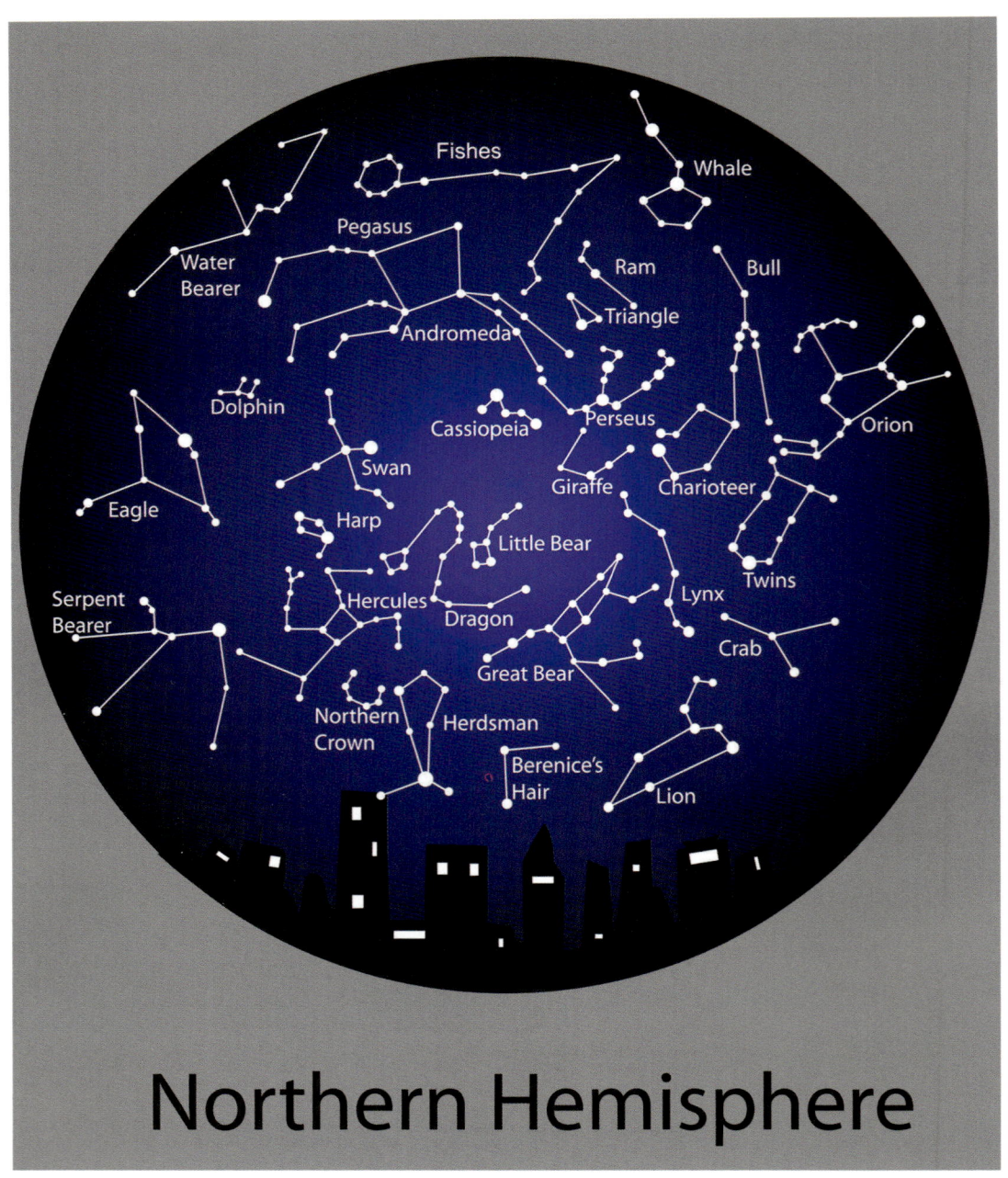

Northern Hemisphere

At any given date, place, and time, completely different constellations are visible. The night sky looks quite different from the Northern or Southern Hemispheres.

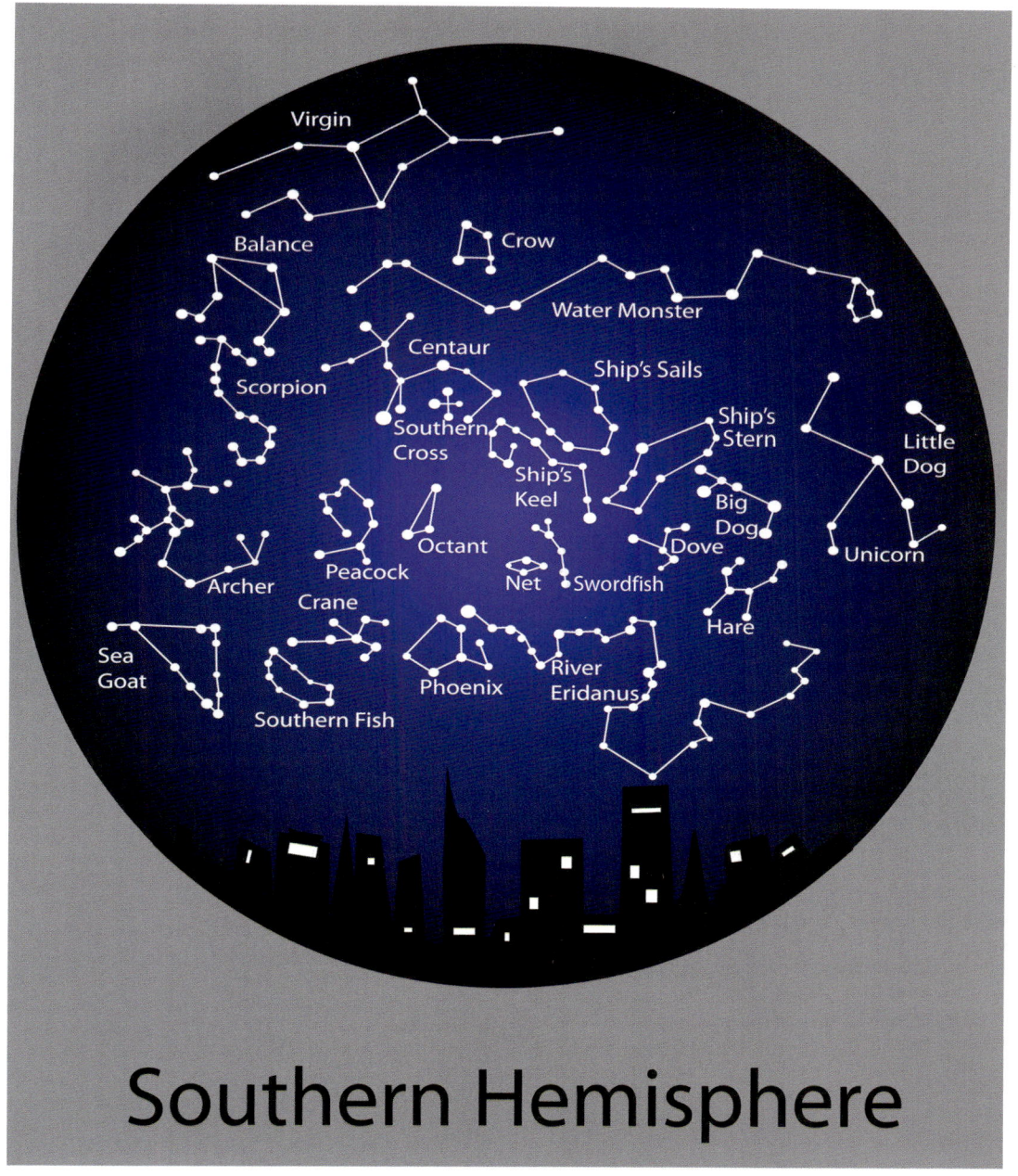

Southern Hemisphere

The Night Sky Changes

Earth's rotation causes the apparent location of constellations to move though the sky each night. And because of Earth's revolution around the sun, some constellations are visible only at certain times of the year.

A few constellations are visible all year long in a certain area of the sky. These are called circumpolar constellations. If you live in the Northern Hemisphere, there are circumpolar constellations that are always visible in the northern sky no matter the season. If you live in the Southern Hemisphere, there are circumpolar constellations that are always visible in the southern sky.

In ancient times, explorers used circumpolar stars and constellations to help with navigation. One of these stars in the Northern Hemisphere, called Polaris (the North Star), never appears to move from its point in the northern sky. Explorers who could locate this star would always know in which direction they were moving.

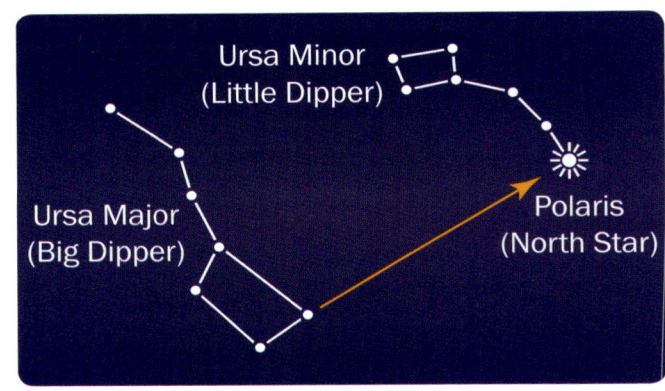

To locate Polaris, look for the constellation called Ursa Minor, also known as the Little Dipper. Polaris is the end point of the dipper's handle. Spotting Ursa Major, the Big Dipper, can help you find Ursa Minor.

Word to Know

Constellations that are always visible in the Northern or Southern Hemispheres no matter the season are *circumpolar*. The prefix *circum-* means "around" and comes from the Latin word *circ*, which means "ring." The word *polar* refers to Earth's North and South Poles.

Gravity

Chapter 7

Suppose you play soccer with friends. You kick the ball high into the air. What happens next? The ball doesn't keep flying into space. It falls back downward after you launch it upward with the force of your kick.

Big Question
What is gravity?

A force is a push or a pull. The force of your kick pushes the ball into upward motion. What must happen to change the direction of its motion and pull the ball down? Give a little more thought to what the direction *down* means. *Down* and *downward* mean toward the ground. But when you investigate how and why objects fall, you can be more specific.

Objects that are not being pushed or held up will fall to Earth's surface.

Earth's Gravity Is a Downward Force

A ball, a book, or any other object dropped from a raised position on Earth falls to the ground. We experience **gravity** as the force that pulls objects toward Earth's center. The ground generally appears flat beneath our feet, and we call that direction "downward." But Earth's shape is a sphere. No matter where one is located on Earth's spherical surface, *downward* always means "in the direction toward the center of the planet."

Vocabulary

gravity, n. a force that pulls objects toward each other

Think About the Center of a Sphere

A sphere is a completely round ball shape. A single point in the center of the sphere is the same distance from all points on its surface.

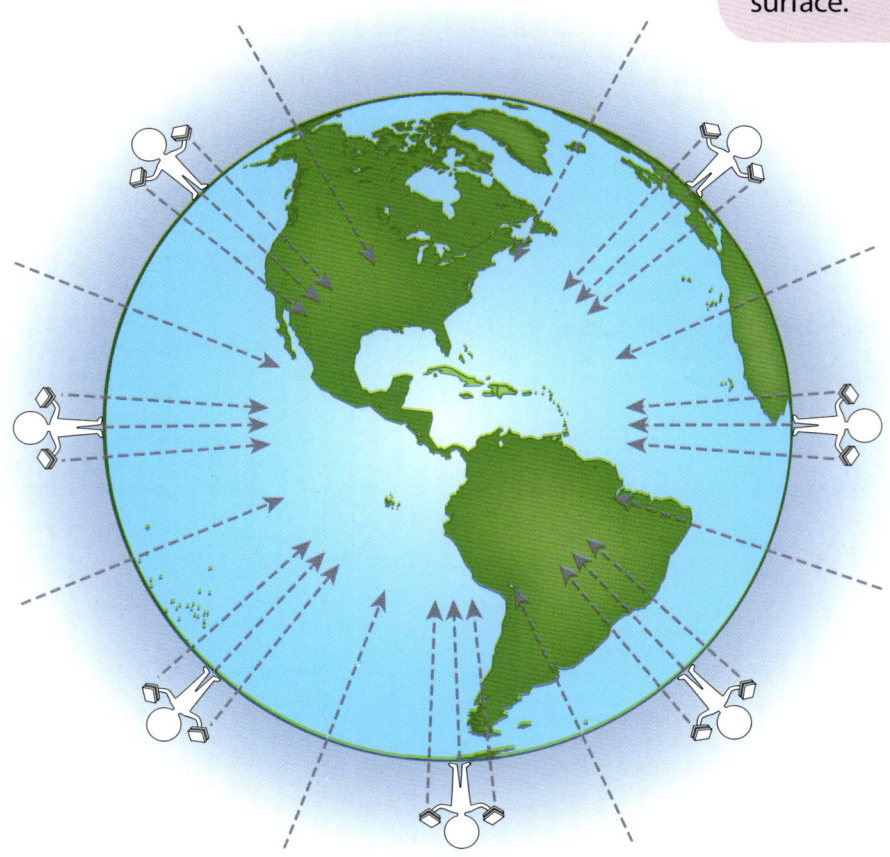

Think About Evidence That Earth Is a Sphere

How do we know Earth is a sphere? Now we can send a spaceship all the way around it. But people knew Earth was spherical long before modern satellites were invented. If so, the Earth must be a sphere.

Lunar eclipses provide possible evidence that Earth is a sphere. A lunar eclipse occurs when the moon passes behind Earth, opposite the sun, and Earth casts a shadow on the moon. Lunar eclipses are visible from different locations on Earth each time they occur. But Earth's shadow always reveals a crescent shape as it moves across the face of the moon. The crescent shape of the shadow's edge is the same no matter which side of Earth is facing the moon during the eclipse. So, Earth seems to be circular from all directions. If you know the Earth rotates, then almost certainly it is most likely a sphere.

Earth's spherical shape is clearly observable from space.

Earth's shape produces a shadow with this same circular curve no matter which part of Earth's surface is blocking the sunlight. So, if the Earth is rotating, it must be a sphere.

All Objects in the Universe Are Affected by Gravity

Gravity Between Earth and the Moon

Objects that are in motion continue the motion in a straight path unless an outside force acts upon them. But the moon's motion around Earth is not a straight line. It follows a nearly, but not perfectly, circular path. Earth's gravitational pull on the moon is what keeps the moon in Earth's orbit.

Gravity exists beyond Earth's surface. It is a force of attraction among all objects in the universe. Gravity governs the motion in the system that includes all objects in space. When any two objects are next to each other, each of them asserts a gravitational force on the other. For example, the moon and Earth are exerting a gravitational force on each other.

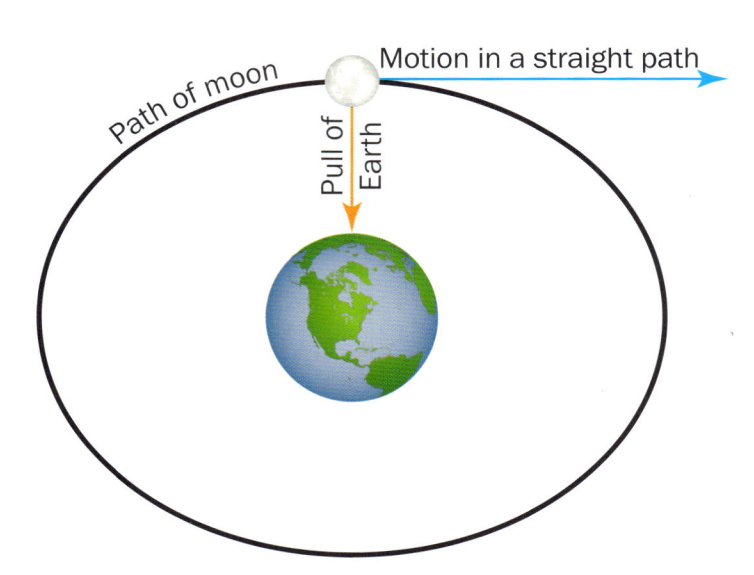

Without Earth's gravitational pull, the moon's motion would be a straight path. A force must act on the moon to change its motion from a straight line to a curved path. That force is Earth's gravity.

At the same time, the moon's gravity also pulls on Earth. Ocean shorelines reveal evidence of the moon's gravitational pull. Tides rise and recede in predictable patterns each day. This pattern is mainly caused by the moon's gravitational force.

As Earth rotates, the moon's gravitational pull is strongest on the side of the planet that faces the moon. Water on that side of Earth bulges toward the moon. This causes high tides along the shoreline. Water on the opposite side of Earth bulges outward, too. So, tides are high on coastlines on that side of the planet at the same time. As the side of Earth with bulging water rotates away from the moon, the water level falls again. This causes low tides. Earth's rotation produces two high tides and two low tides in most coastal locations each day.

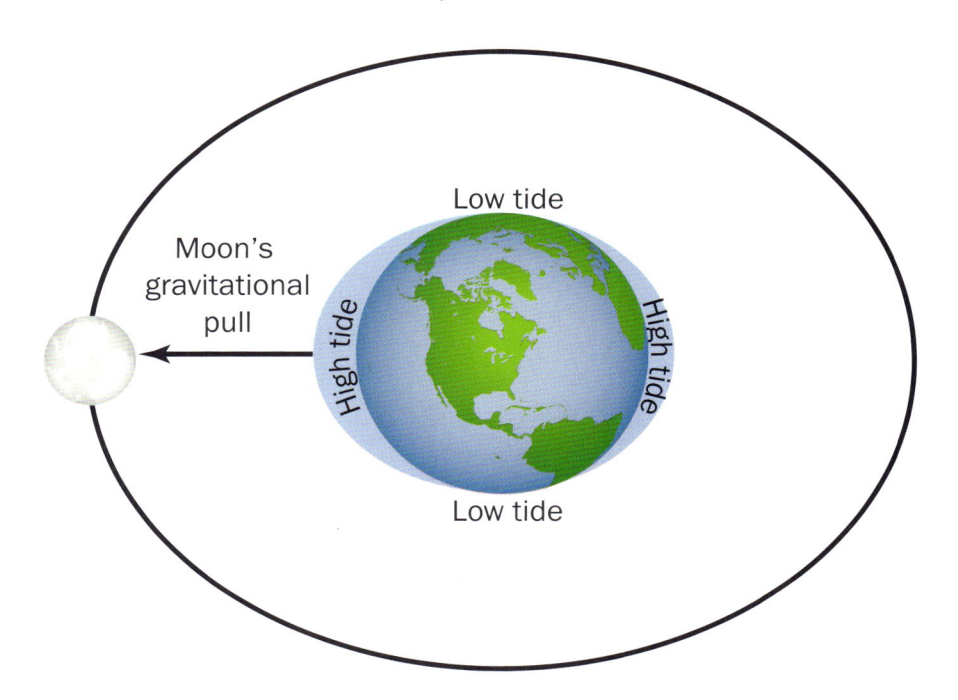

Gravity Between the Sun and the Planets

Understanding gravity helps scientists explain how our solar system was formed. Our solar system formed over four billion years ago as a spinning cloud of gases and dust. Gravitational forces among the particles and gases pulled them toward one another. The sun formed at the center of the cloud. Materials farther out pulled together to form planets.

The more matter there is in an object—the more mass it has—the greater its gravitational pull. The sun's huge mass means its gravitational force affects every object in our solar system, including planets, moons, and asteroids. All the objects in the solar system that formed around the sun are still revolving around the sun today. The speed of the planets' motion and the sun's gravitational pull on them keeps them on a nearly circular path.

The sun's gravitational pull keeps other objects in the solar system in orbit around it. Planets are massive enough to keep objects very near to them in their own orbits.

Women and Studying Space

Chapter 8

When you think of the size of the solar system, it is amazing that we know as much as we do about it! The National Aeronautics and Space Administration, or NASA, was created in 1958. Since that time, twelve American astronauts have walked on the moon's surface. Hundreds of others have participated in space flights. And about two thousand spacecrafts have been sent into Earth's orbit and beyond. These missions have all been supported by countless scientists and engineers on the ground. Many of these investigators were and are women.

Big Question

How have people learned about space systems?

Some astronauts, though not all, work as pilots earlier in their careers.

Women Astronauts

About fifty American women have flown in various NASA space missions. However, throughout the years, women from other countries, including Russia and China, have flown to space, too. Space programs from different countries often work together to help deepen our understanding of Earth, the moon, and the solar system.

Valentina Tereshkova

On June 16, 1963, a Russian cosmonaut named Valentina Tereshkova became the first woman in space. She was just twenty-six years old and flew aboard a spacecraft known as *Vostok 6*. During the mission, Tereshkova completed forty-five orbits of Earth over a period of seventy hours and fifty minutes. The Soviet space program chose Tereshkova because she was an excellent parachute jumper. Her parachuting skills came in handy once the *Vostok 6* reentered Earth's atmosphere, when Tereshkova parachuted from the spacecraft and landed back on Earth. She went on to graduate from a Russian Air Force engineering academy and receive a degree in technical science.

Valentina Tereshkova

Sally Ride

Just over twenty years after Valentina Tereshkova's historic mission, Sally Ride became the first American woman in space. At the age of thirty-two, she was the youngest American to ever travel to space at that time.

Ride was born May 26, 1951, in Encino, California. She pursued college degrees in both English and physics. She then went on to earn a master's degree and a PhD in physics from Stanford University.

After being selected by NASA to be part of the 1978 astronaut class, Ride began her training. She launched into space aboard the *Challenger* STS-7 shuttle on June 18, 1983. She had many different responsibilities on the mission, including launching communications satellites, operating specific parts of the shuttle, and conducting experiments. After one week, the mission was complete. Ride spent eight days observing Earth from space on her second mission in 1984. Her career spanned decades and included both research and teaching.

Shortly after her death in 2012, President Barack Obama awarded Ride a Presidential Medal of Freedom.

Sally Ride spent six days in space aboard her first *Challenger* mission.

Mae Jemison

Mae Jemison was born in 1956 in Decatur, Alabama. She enrolled at Stanford University at just sixteen years of age! She graduated with degrees in both chemical engineering and African American studies before going on to become a medical doctor. Jemison practiced medicine in Cambodia, in West Africa, and in the United States before she was selected by NASA for astronaut training.

Jemison journeyed into Earth orbit aboard the shuttle *Endeavour* in 1992. She was the first African-American woman to travel to space. As a physician, Dr. Jemison studied the affects of zero gravity on the human body during space travel. She and the *Endeavour* crew aboard the shuttle mission STS-47 orbited Earth for nearly eight days.

Mae Jemison retired from NASA in 1993 and has since worked on many projects involving health care, education, technology, and research. She has also written several books for young readers.

Mae Jemison began training for her mission with NASA in 1987.

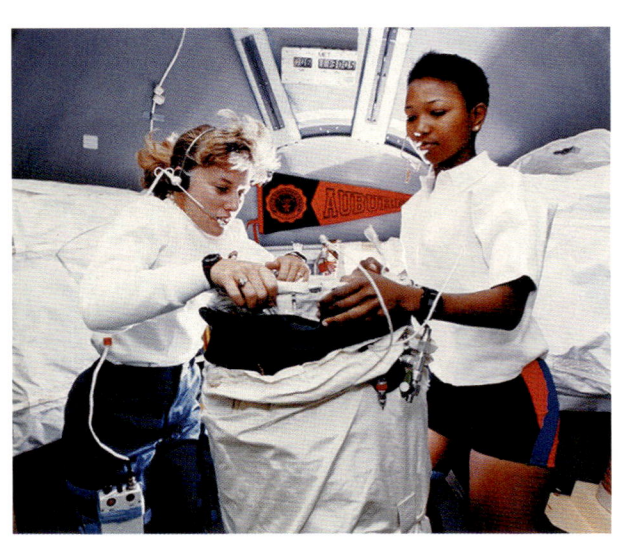

Jan Davis, who was aboard STS-47 with Mae Jemison, flew on additional space shuttle missions. Davis is a mechanical engineer.

Women Scientists and Engineers

Nancy Grace Roman

Throughout this unit, you have seen several photos from the Hubble Space Telescope. One of the women who made that project a reality was scientist Nancy Grace Roman. Born in 1925, Roman grew up during a time when women were not encouraged to pursue educations in science and math. But Roman loved astronomy from the time she was a child and wanted to study space. She later discovered a star, helped plan and develop the Hubble Space Telescope, and served as director of the Astronomical Data Center. Roman was also NASA's first female executive.

Nancy G. Roman worked for NASA for over twenty years.

Nancy G. Roman is pictured here with Buzz Aldrin, the second astronaut to set foot on the moon.

Katherine Johnson

Katherine Johnson was born in West Virginia in 1918. As a child, Johnson loved math. In fact, her math skills were so advanced that she skipped several grades in school and attended classes at a college campus by the time she was just thirteen years old. After receiving a degree in math, Johnson became a teacher for a short time before deciding to accept a job with the organization that would later become NASA. At that time, computers like the ones we have today did not yet exist, and people were relied upon to solve difficult math problems, sometimes quickly. Johnson's advanced equations and calculations made it possible for astronaut John Glenn to orbit Earth for the first time in 1962. Her calculations also helped NASA send astronauts to the moon.

Much of Katherine Johnson's work related to determining the positions of objects in orbit.

Many decades after her work for NASA, Katherine Johnson was awarded the Presidential Medal of Freedom in 2015.

Glossary

A

absolute brightness, n. how bright a star is from a standard distance of 32.6 light-years (**25**)

apparent brightness, n. how bright a star appears compared with other stars when all are viewed from Earth (**25**)

axis, n. an imaginary line through the center of an object that is a fixed point of reference (**16**)

C

constellation, n. a group of stars that forms a recognizable pattern in the sky (**27**)

G

galaxy, n. a collection of stars, their solar systems, dust, and gas (**7**)

gravity, n. a force that pulls objects toward each other (**32**)

L

light-year, n. the distance that light travels in one Earth year (**25**)

lunar eclipse, n. an event during which the moon passes directly behind Earth and into its shadow (**21**)

M

moon phase, n. a stage in the repeating, predictable pattern of change in the moon's appearance from Earth (**20**)

O

orbit, n. the oval-shaped path an object follows as it revolves around another object in space (**v.** to revolve around another object) (**1**)

S

solar eclipse, n. an event during which the moon's shadow blocks all or some of the sun's light from reaching Earth (**22**)

solar system, n. a system of objects in space that includes at least one star, planets, their moons, asteroids, comets, and other space debris (**1**)

star, n. a space object that gives off its own heat and light (**23**)

U

universe, n. all of the existing matter and energy in space (**9**)

Core Knowledge®

CKSci™
Core Knowledge SCIENCE™

Series Editor-in-Chief
E.D. Hirsch Jr.

Editorial Directors
Daniel H. Franck and Richard B. Talbot

Subject Matter Expert

Charles Tolbert, PhD
Professor Emeritus
Department of Astronomy
University of Virginia
Charlottesville, VA

Illustrations and Photo Credits

AB Forces News Collection / Alamy Stock Photo: 37
Alan Dyer / VWPics / Alamy Stock Photo: 22a, 24a
Alexander Aldatov / Alamy Stock Photo: 24c
Allexxandar / Alamy Stock Photo: 11
Alpha Historica / Alamy Stock Photo: 42a
Andrey Armyagov / Alamy Stock Photo: 13
Angelina Stoykova / Alamy Stock Vector: Cover C, 27
arkela / Alamy Stock Photo: 28-29
B.A.E. Inc. / Alamy Stock Photo: 33a
Dominic Gentilcore / Alamy Stock Photo: 8b
dotted zebra / Alamy Stock Photo: 3b, 7a
dpa picture alliance / Alamy Stock Photo: 42b
Elvele Images Limited / Alamy Stock Photo: 5a, 5b
Everett Collection / SuperStock: Cover B, 40a, 40b
Gado Images / Alamy Stock Photo: 39
Golden Pixels / Purestock / SuperStock: 31
Irina Dmitrienko / Alamy Stock Photo: Cover A, 4a, 9
Jerome Murray - CC / Alamy Stock Photo: 17a
Lee Dalton / Alamy Stock Photo: 3a
Luc Novovitch / Alamy Stock Photo: i, iii, 10
Mike Kemp / Blend Images / SuperStock: 15
NASA Image Collection / Alamy Stock Photo: 2b, 6
NASA Photo / Alamy Stock Photo: 12
Nerthuz / Alamy Stock Photo: 2a
Olekcii Mach / Alamy Stock Photo: 1
Science and Society / SuperStock: 21
Science History Images / Alamy Stock Photo: 4b, 7b, 24b, 41a, 41b
Science Photo Library / Alamy Stock Photo: 23
Science Photo Library / SuperStock: 36
SPUTNIK / Alamy Stock Photo: 38
Steppeland / Alamy Stock Photo: 33b
Stocktrek Images, Inc. / Alamy Stock Photo: 8a, 14, 25, 26
Universal Images Group North America LLC / DeAgostini / Alamy Stock Photo: 18
Westend61 GmbH / Alamy Stock Photo: 19
World History Archive / Alamy Stock Photo: Cover D, 22b